GROWING LEAF AMARANTH

FOR BUSINESS

Complete Beginners Guide To Understand And Master How To Grow Leaf Amaranth From Scratch (Cultivation, Care, Management, Harvest, Profit And More)

HARRISON DAMON

DISCLAIMER

The information provided in this book, is intended for informational purposes only. The author makes no representations or warranties of any kind, express or implied, about the completeness, accuracy, reliability, suitability, or availability of the information contained in this book. Any reliance you place on such information is strictly at your own risk.

The author shall not be liable for any loss or damage arising from the use of this book or the information contained herein. It is your responsibility to conduct thorough research and seek professional advice before making any decisions based on the content of this book.

This book may contain references or mentions of individuals, products, websites, organizations, or other names for illustrative purposes only. The author does not endorse, recommend, or have any affiliation with the mentioned entities unless explicitly stated

Table of Contents

CHAPTER ONE

INTRODUCTION

Background And Significance

Amaranth, a versatile and nutrition-packed leafy green, has gained prominence in recent years due to its exceptional health benefits and culinary versatility. The plant, known for its vibrant green leaves, is not only a source of essential nutrients but also a promising business opportunity for agricultural enthusiasts. Understanding the background and significance of growing leaf amaranth is crucial for those looking to tap into this thriving market.

Historical Context

Amaranth traces its roots back to ancient civilizations, where it was revered for its nutritional value and easy cultivation. In modern times, its resurgence is driven by a growing awareness of the need for nutrient-dense foods and sustainable farming practices. Exploring the historical context of amaranth sets the stage for understanding its timeless appeal and adaptability.

Nutritional Powerhouse

Leaf amaranth stands out as a nutritional powerhouse, packed with vitamins, minerals, and antioxidants. Rich in iron, calcium, and essential amino acids, it addresses malnutrition concerns and contributes to overall well-being. As consumers increasingly prioritize health-conscious choices, the demand for nutrient-dense greens like amaranth continues to rise, making it a lucrative venture for aspiring agripreneurs.

Objectives of the Book

To embark on a successful journey of growing leaf amaranth for business, it's essential to have a clear set of objectives. This book is crafted with specific goals in mind, catering to both beginners and experienced farmers aiming to capitalize on the burgeoning amaranth market.

Comprehensive Cultivation Guidance

One of the primary objectives of this book is to provide a step-by-step guide to cultivating leaf amaranth. From selecting the right variety and preparing the soil to managing pests and diseases, readers will gain comprehensive insights into the entire cultivation process. The goal is to empower individuals with the knowledge and skills needed to foster a thriving amaranth farm.

Market Analysis and Trends

Understanding market dynamics is crucial for any business venture. This book delves into market analysis, shedding light on current trends, consumer preferences, and potential challenges within the amaranth industry. By equipping readers with market intelligence, they can make informed decisions, positioning their amaranth business for success in a competitive landscape.

Sustainable Practices

The book aims to promote sustainable farming practices in amaranth cultivation. From eco-friendly pest control methods to water conservation strategies, emphasis is placed on minimizing environmental impact. Integrating sustainability into the farming process not only aligns with global trends but also enhances the market appeal of the produced amaranth.

Target Audience

Identifying the right audience is crucial for the book's effectiveness in imparting valuable knowledge. The target audience encompasses a diverse group of individuals, each with unique interests and aspirations related to leaf amaranth cultivation.

Aspiring Agripreneurs

For individuals looking to venture into agriculture as a business, this book serves as a comprehensive resource. It provides the foundational knowledge needed to start and sustain a successful amaranth farming enterprise. Whether someone is new to farming or transitioning from other crops, the book caters to their entrepreneurial aspirations.

Experienced Farmers Seeking Diversification

Experienced farmers seeking to diversify their crops and capitalize on emerging trends will find this book valuable. By offering insights into the intricacies of amaranth cultivation and its market potential, it equips seasoned farmers with the tools to expand their portfolios and stay ahead in a dynamic agricultural landscape.

Health and Nutrition Enthusiasts

Individuals passionate about health and nutrition form another segment of the target audience. The book's focus on amaranth's nutritional benefits and its role in promoting well-being appeals to those seeking to incorporate nutrient-dense foods into their diet or exploring business opportunities aligned with health-conscious consumer trends.

In conclusion, "Growing Leaf Amaranth for Business" is more than just a guide—it's a comprehensive resource that explores the historical significance, outlines specific objectives, and caters to a diverse audience eager to embrace the potential of leaf amaranth cultivation as a thriving business opportunity.

CHAPTER TWO

UNDERSTANDING LEAF AMARANTH

Botanical Overview

Leaf amaranth, scientifically known as Amaranthus spp., belongs to the Amaranthaceae family. This versatile and hardy plant is recognized for its vibrant green leaves and various uses in culinary and traditional medicine. Amaranth plants typically grow as annuals, exhibiting a rapid growth cycle, making them suitable for year-round cultivation. The plant's distinctive features include erect stems, ovate to lanceolate leaves, and small, inconspicuous flowers. Understanding the botanical characteristics of leaf amaranth is crucial for successful cultivation and commercialization.

Leaf amaranth thrives in diverse climates, displaying adaptability to both tropical and subtropical regions. It prefers well-drained soils with a slightly acidic to neutral pH. The plant's resilience against pests and diseases contributes to its popularity among farmers.

Successful cultivation requires adequate sunlight and regular watering, ensuring optimal growth conditions. Being a prolific seed producer, leaf amaranth offers a sustainable source for continuous harvests, enhancing its appeal for commercial cultivation.

<u>Varieties of Leaf Amaranth</u>

Diversity in leaf amaranth varieties opens up opportunities for farmers to choose the most suitable cultivars based on their specific needs and growing conditions. Some popular varieties include:

1. Red Leaf Amaranth (Amaranthus tricolor)

This variety is renowned for its striking reddish-purple leaves, adding aesthetic appeal to culinary dishes. Red leaf amaranth is not only visually appealing but also rich in nutrients, making it a preferred choice for both home gardens and commercial cultivation.

2. Green Leaf Amaranth (Amaranthus viridis)

Characterized by its vibrant green leaves, this variety is a staple in many cuisines globally. It boasts a mild flavor, making it versatile for various culinary applications. Green leaf amaranth is known for its quick growth and adaptability to different environmental conditions.

3. Chinese Spinach (Amaranthus tricolor)

Also known as bayan in some regions, Chinese spinach is a popular leaf amaranth variety with tender leaves and a slightly tangy flavor. Its rapid growth and high yield make it a favored choice for commercial cultivation.

Understanding the distinct features and growth patterns of these varieties empowers farmers to make informed decisions, maximizing their yield and meeting market demands effectively.

Nutritional Value

Leaf amaranth stands out not only for its agronomic benefits but also for its exceptional nutritional profile.

The leaves are a rich source of essential vitamins, minerals, and antioxidants. Key nutritional components include:

1. Vitamins

Leaf amaranth is a powerhouse of vitamins, particularly vitamin A, vitamin C, and folate. These vitamins play crucial roles in supporting immune function, promoting healthy skin, and aiding in cell division and growth.

2. Minerals

The plant is a good source of essential minerals such as iron, calcium, and potassium. Incorporating leaf amaranth into a balanced diet helps in maintaining bone health, regulating blood pressure, and preventing iron deficiency anemia.

3. Antioxidants

The presence of antioxidants, including flavonoids and polyphenols, contributes to the plant's anti-inflammatory properties. These compounds help combat oxidative stress in the body, potentially reducing the risk of chronic diseases.

Understanding the nutritional benefits of leaf amaranth positions it as a valuable addition to a healthy diet, driving consumer interest and market demand.

Economic Potential

The economic potential of growing leaf amaranth for business is substantial, thanks to its versatile applications and increasing consumer awareness of its nutritional benefits. Key factors contributing to its economic viability include:

1. Market Demand

The rising demand for fresh and nutritious greens in both local and international markets creates a favorable environment for leaf amaranth cultivation. As consumers prioritize health and wellness, the market for nutrient-rich leafy greens continues to expand.

2. Culinary Uses

Leaf amaranth's popularity in various cuisines, especially in Asian and African dishes, positions it as a sought-after ingredient in the culinary world. Restaurants, food processors, and culinary enthusiasts contribute to the demand for high-quality leaf amaranth.

3. Sustainable Farming

Leaf amaranth's ability to thrive in diverse climates and its resilience to pests and diseases make it an attractive option for sustainable farming practices. Its quick growth cycle and prolific seed production contribute to consistent yields, reducing production risks for farmers.

In conclusion, growing leaf amaranth for business requires a comprehensive understanding of its botanical characteristics, varieties, nutritional value, and economic potential. By leveraging this knowledge, farmers can optimize cultivation practices, meet market demands, and capitalize on the economic opportunities offered by this versatile leafy green.

CHAPTER THREE

MARKET ANALYSIS

Demand and Supply Trends

Overview: The demand for leaf amaranth has witnessed a significant surge in recent years, driven by a growing awareness of its nutritional benefits and versatile culinary applications. This leafy green, rich in essential vitamins and minerals, has found favor among health-conscious consumers and those seeking diverse and exotic ingredients. As a result, the market for growing leaf amaranth is experiencing a steady upward trajectory.

Growing Demand: One of the key drivers of demand is the increasing interest in plant-based diets. Consumers are actively seeking nutritious alternatives to traditional vegetables, and leaf amaranth's high nutritional content positions it as an attractive option. Furthermore, the rise in global population, coupled with a focus on sustainable agriculture, has spurred demand for crops that offer high yields with minimal environmental impact.

Supply Dynamics: The supply side has responded to the heightened demand for leaf amaranth, with more farmers and agricultural businesses recognizing its economic potential. Advances in farming techniques, including hydroponics and vertical farming, have also contributed to increased production efficiency. However, challenges such as seasonal variations and susceptibility to certain pests underscore the need for sustainable farming practices to ensure a consistent supply.

Consumer Preferences

Nutritional Appeal: Leaf amaranth has become a sought-after choice among health-conscious consumers due to its impressive nutritional profile. Rich in vitamins A and C, calcium, iron, and protein, it aligns with the growing trend of incorporating nutrient-dense foods into daily diets. The leaf's versatility in various culinary applications, from salads to smoothies, further enhances its appeal.

Culinary Versatility: Consumers appreciate the adaptability of leaf amaranth in diverse culinary styles. Its mild flavor and tender texture make it suitable for raw consumption in salads or as a cooked component in soups, stir-fries, and side dishes. As global palates continue to evolve, the culinary flexibility of leaf amaranth positions it as a star ingredient in various cuisines.

Sustainability Consciousness: A rising trend in consumer preferences is the emphasis on sustainability. Leaf amaranth, with its relatively low resource requirements compared to traditional crops, aligns with the values of environmentally conscious consumers. As awareness about sustainable farming practices grows, businesses involved in growing leaf amaranth can leverage this aspect to build a positive brand image.

Competitor Landscape

Emerging Players: The landscape of leaf amaranth cultivation has seen the emergence of new players, ranging from small-scale local farmers to larger agribusinesses. As the market expands, competition is intensifying, prompting growers to differentiate themselves based on factors such as quality, organic cultivation practices, and innovative packaging.

Technological Advancements: In the competitive arena, technological advancements play a crucial role. Precision farming, data analytics, and automation are being employed to enhance yield, reduce production costs, and ensure consistent quality. Companies investing in research and development to optimize cultivation processes are gaining a competitive edge, contributing to the overall sophistication of the industry.

Global and Local Dynamics: The global nature of the leaf amaranth market means that businesses must navigate both local and international dynamics. Local farmers may benefit from regional demand surges, while larger enterprises may explore export opportunities. Understanding the intricacies of regional preferences, climate variations, and distribution channels is vital for businesses aiming to thrive in the competitive landscape.

Potential Markets

Emerging Economies: The potential for growth in leaf amaranth cultivation is particularly promising in emerging economies. Rapid urbanization, increasing disposable incomes, and a shift towards healthier lifestyles are creating new markets for nutritious and exotic produce. Identifying and strategically entering these markets can be a lucrative opportunity for businesses looking to expand their footprint.

Health and Wellness Markets: Leaf amaranth aligns with the booming health and wellness industry, providing an opportunity to tap into markets specifically focused on organic, natural, and functional foods. Collaborations with health food stores, fitness centers, and wellness retreats can be avenues for businesses to position themselves as key players in this niche segment.

Culinary Tourism and Foodservice Industry: As culinary tourism gains popularity, there is a growing demand for unique and exotic ingredients. The food service industry, including restaurants, cafes, and catering services, presents a lucrative market for leaf amaranth. Collaborations with chefs and influencers can elevate the visibility of this leafy green, making it a staple in both home kitchens and professional culinary settings.

In conclusion, growing leaf amaranth for business holds immense potential, driven by evolving consumer preferences, competitive dynamics, and untapped markets. Navigating this landscape requires a strategic approach, incorporating sustainable practices, technological advancements, and a keen understanding of both local and global market trends. Businesses that adeptly position themselves in response to these factors are well-positioned to capitalize on the flourishing market for leaf amaranth.

CHAPTER FOUR

SITE SELECTION AND PREPARATION

Soil Requirements:

Growing leaf amaranth for business requires careful consideration of soil conditions to ensure optimal growth and yield. Amaranth thrives in well-drained, fertile soils with a slightly acidic to neutral pH ranging between 6.0 and 7.5. Before initiating cultivation, conducting a soil test is crucial to assess nutrient levels and soil composition. This information will guide the application of necessary soil amendments to create an ideal environment for amaranth cultivation.

Amaranth is particularly responsive to organic matter, so incorporating well-rotted compost or manure into the soil enhances its structure and nutrient content. Additionally, ensuring good aeration is vital for healthy root development. Adequate drainage is essential to prevent waterlogged conditions, which can lead to root rot and other diseases. Soil preparation should aim to create a balanced and nutrient-rich foundation for the amaranth crop.

2. Climate Considerations:

Choosing the right location for growing leaf amaranth involves careful consideration of the climate. Amaranth is a warm-season crop that thrives in temperatures between 65°F to 90°F (18°C to 32°C). It is sensitive to frost, so planting should occur after the last expected frost in the spring. The climate should also provide a growing season long enough for amaranth to reach maturity before adverse weather conditions, such as extreme heat or cold, set in.

Amaranth requires full sunlight for optimum growth, so selecting a site with maximum exposure to sunlight is essential. Adequate sunlight ensures robust photosynthesis, leading to healthy foliage and abundant yields. Understanding the specific climatic requirements of amaranth is fundamental to site selection for successful commercial cultivation.

3. Land Preparation Techniques:

Efficient land preparation is a critical factor in the success of growing leaf amaranth for business. Begin by clearing the selected site of any debris, rocks, or unwanted vegetation. This ensures a clean and uniform planting area. Subsequent plowing or tilling should be done to break up the soil, facilitating root penetration and water absorption.

Incorporate soil amendments based on the results of the soil test, focusing on enhancing fertility and structure. Well-prepared beds or rows with proper spacing between plants contribute to efficient cultivation and harvesting. Implementing raised beds can aid in drainage, especially in areas prone to waterlogging. Mulching the soil surface helps retain moisture, suppress weeds, and regulate soil temperature.

Precision in land preparation not only ensures an optimal environment for amaranth growth but also facilitates subsequent farm operations, contributing to overall efficiency in the cultivation process.

4. Infrastructure Setup:

Establishing the necessary infrastructure is a pivotal aspect of growing leaf amaranth for business. Adequate irrigation systems, such as drip or sprinkler systems, should be installed to ensure consistent moisture levels in the soil. Efficient water management is crucial for preventing drought stress and maximizing yield.

In regions with unpredictable weather patterns, the installation of protective structures such as high tunnels or greenhouses can provide a controlled environment for amaranth cultivation. This is particularly beneficial in extending the growing season and protecting the crop from adverse weather conditions.

Furthermore, consider implementing a nutrient management plan, incorporating fertilization schedules and pest control measures. Implementing a trellising system can support the upright growth habit of certain amaranth varieties, facilitating easier harvesting and better air circulation.

Investing in quality harvesting and post-harvest handling equipment is essential for preserving the quality of the harvested amaranth. This includes proper packaging facilities and storage infrastructure to maintain freshness and market appeal.

In conclusion, the successful cultivation of leaf amaranth for business demands a holistic approach, encompassing soil preparation, climate considerations, land preparation techniques, and the establishment of essential infrastructure. A well-planned and executed strategy in each of these aspects is vital for achieving a thriving amaranth crop with optimal yield and quality for commercial purposes.

CHAPTER FIVE

SEED SELECTION AND GERMINATION

Choosing Quality Seeds

Choosing quality seeds is a critical first step in the successful cultivation of leaf amaranth. Quality seeds are essential for ensuring a robust and healthy crop. When selecting seeds, it is crucial to consider factors such as seed purity, germination rate, and disease resistance. Opt for certified seeds from reputable suppliers to ensure that you are starting with the best possible genetic material.

Seed purity is vital to maintain the desired characteristics of the leaf amaranth variety you intend to grow. Inspect the seeds for any signs of impurities, off-types, or contaminants. It is advisable to purchase seeds from trusted sources that provide detailed information on the genetic purity of their seed stock.

Germination rate is another crucial factor in seed selection. Choose seeds with a high germination rate to ensure a uniform stand and maximize the overall yield. Conducting a germination test before planting can help you assess the viability of the seeds and identify any potential issues.

Consider the disease resistance of the selected seed variety. Opting for seeds with resistance to common pests and diseases can reduce the need for chemical interventions later in the growing process, promoting a more sustainable and eco-friendly approach to cultivation.

Seed Treatment

Seed treatment plays a pivotal role in enhancing the germination and early growth of leaf amaranth. Proper seed treatment can improve seedling vigor, protect against soil-borne pathogens, and enhance nutrient absorption. There are various seed treatment methods, including priming, coating, and biological treatments.

Priming involves soaking the seeds in water or a nutrient solution to initiate the germination process. This method can improve the uniformity and speed of germination, leading to more robust seedlings. Additionally, priming helps the seeds withstand environmental stress factors, such as drought or temperature fluctuations.

Coating seeds with a protective layer can offer several benefits. Coated seeds may be easier to handle, have improved resistance to pests and diseases, and provide a more controlled release of nutrients during germination. This method is particularly useful in challenging growing conditions.

Biological seed treatments involve applying beneficial microorganisms to the seeds, promoting symbiotic relationships that enhance plant health. These microorganisms can protect the seedlings from harmful pathogens and improve nutrient uptake, contributing to overall crop resilience.

Germination Techniques

Successful germination is the foundation of a healthy leaf amaranth crop. Employing effective germination techniques is crucial to ensure a high germination rate and uniform seedling emergence. Common germination techniques include direct sowing, seedbeds, and hydroponic systems.

Direct sowing involves planting seeds directly in the field where the crop will grow. This method is straightforward and suitable for leaf amaranth, provided that soil conditions are optimal. Ensure proper soil preparation, including adequate moisture and nutrient levels, to support germination.

Seedbeds are controlled environments where seeds are germinated before being transplanted to the main field. This technique allows for better monitoring of environmental conditions, protection against adverse weather, and easier management of seedlings. Use well-prepared seedbeds with a suitable growing medium to facilitate germination.

Hydroponic systems offer a soil-less alternative for germinating leaf amaranth seeds. This technique involves growing plants in nutrient-rich water solutions. Hydroponics can provide precise control over environmental factors, leading to faster and more efficient germination. However, it requires specialized equipment and expertise.

Nursery Management

Proper nursery management is essential for fostering healthy seedlings before transplanting them into the main field. Key aspects of nursery management include seedling care, irrigation, and disease control.

Ensure optimal conditions for seedling growth by providing adequate light, temperature, and humidity in the nursery. Monitor the seedlings regularly to identify and address any issues promptly. Proper spacing between seedlings is crucial to prevent overcrowding, which can lead to competition for resources.

Implement a consistent and well-regulated irrigation schedule to prevent both under-watering and over-watering. The soil or growing medium should have good water retention properties to support the development of robust root systems. Avoid waterlogged conditions, as they can lead to diseases and root rot.

Implement disease control measures in the nursery to prevent the spread of pathogens. This includes practicing good hygiene, using disease-resistant varieties, and applying appropriate fungicides or biopesticides. Regular scouting for signs of diseases and pests is crucial for early intervention and prevention of widespread issues.

In conclusion, the successful cultivation of leaf amaranth begins with selecting high-quality seeds, employing effective seed treatment methods, implementing proper germination techniques, and managing the nursery environment diligently. By focusing on these key aspects, farmers can establish a strong foundation for a productive and healthy leaf amaranth crop.

CHAPTER SIX

CROP MANAGEMENT

Planting Procedures

Site Selection:

Selecting the right site for cultivating leaf amaranth is crucial for a successful business venture. Choose a location with well-drained soil that is rich in organic matter. Amaranth thrives in full sunlight, so ensure the chosen site receives ample sunlight throughout the day. Conduct a soil test to determine its pH and nutrient levels, making necessary amendments for optimal growth.

Seed Selection and Sowing:

Carefully choose high-quality amaranth seeds from reputable suppliers. The selection process should consider factors such as disease resistance, yield potential, and adaptability to local climate conditions. Sow the seeds directly into the prepared soil at the recommended depth, spacing, and density. Adequate spacing ensures proper air circulation and reduces the risk of disease.

Germination and Thinning:

Maintain consistent moisture levels to facilitate uniform germination. Once the seedlings emerge, thin them to the recommended spacing to avoid overcrowding. Thinning promotes stronger and healthier plants by reducing competition for nutrients and sunlight. Monitor the germination process closely, as amaranth is sensitive to water stress during this crucial stage.

Irrigation Practices

Drip Irrigation Systems:

Efficient water management is essential for leaf amaranth cultivation. Implementing drip irrigation systems is highly recommended, as they deliver water directly to the base of the plants, minimizing water wastage and reducing the risk of fungal diseases. Drip systems also help maintain consistent soil moisture levels, crucial for optimal growth and nutrient absorption.

Watering Schedule:

Establish a regular watering schedule, considering the specific water needs of amaranth at different growth stages. Adequate moisture is critical during germination, establishment, and flowering. Be mindful of environmental factors such as temperature and humidity, adjusting the watering frequency accordingly. Avoid overwatering, as excessive moisture can lead to root diseases.

Mulching Techniques:

Apply a layer of organic mulch around the amaranth plants to conserve soil moisture and suppress weed growth. Mulching helps regulate soil temperature, preventing extremes that may stress the plants. Organic mulch also contributes to soil fertility as it decomposes over time. Consider using materials like straw or shredded leaves for effective mulching.

Fertilization Strategies

Soil Analysis:

Conduct regular soil tests to assess nutrient levels and pH. Amaranth plants have specific nutritional requirements at different growth stages. Based on the soil analysis, tailor a fertilization plan to meet these needs. Utilize organic fertilizers or well-balanced synthetic fertilizers to provide essential nutrients for robust plant development.

Nitrogen Management:

Amaranth is known for its high nitrogen demand, particularly during the vegetative growth phase. Incorporate nitrogen-rich fertilizers at the appropriate times to enhance leaf development. However, avoid excessive nitrogen, as it may lead to delayed flowering and reduced seed production. Striking the right balance is crucial for achieving optimal yield and quality.

Foliar Feeding:

Supplement traditional soil fertilization with foliar feeding to ensure efficient nutrient absorption. Applying a nutrient-rich solution directly to the leaves allows for rapid nutrient uptake, addressing deficiencies promptly. Select a balanced foliar fertilizer and administer it during the early morning or late afternoon to maximize absorption.

Pest and Disease Management
Integrated Pest Management (IPM):

Implement an integrated pest management approach to control pests effectively while minimizing environmental impact. Identify common amaranth pests, such as aphids and flea beetles, and employ natural predators or biopesticides. Regularly inspect the crop for signs of infestation and intervene promptly to prevent widespread damage.

Disease Prevention: Preventive measures are key to managing diseases in leaf amaranth. Practice crop rotation to break disease cycles and reduce the buildup of soilborne pathogens. Ensure proper spacing between plants for air circulation, as this helps minimize conditions conducive to fungal infections. Use disease-resistant varieties whenever possible to mitigate the risk of crop losses.

Early Detection and Treatment:

Monitor the crop regularly for any signs of pest or disease infestation. Early detection allows for timely intervention, preventing the escalation of problems. In case of disease outbreaks, employ appropriate fungicides or pesticides with minimal impact on beneficial organisms. Follow recommended application rates and schedules to achieve effective control.

In conclusion, successful leaf amaranth cultivation for business involves meticulous planning and adherence to best practices in planting, irrigation, fertilization, and pest and disease management. By implementing these strategies, farmers can optimize yield, quality, and overall profitability in the competitive leaf amaranth market.

CHAPTER SEVEN

GROWTH STAGES AND MONITORING

Understanding Growth Phases

Leaf Amaranth, a nutritious and versatile leafy green, undergoes distinct growth phases that are crucial for successful cultivation. Understanding these phases is vital for optimizing yields and ensuring a thriving amaranth crop.

Germination and Seedling Stage

The journey of growing leaf amaranth begins with germination. During this phase, seeds absorb water, swell, and initiate the process of sprouting. Once germinated, seedlings emerge, and the first true leaves appear. Adequate soil moisture and warmth are critical factors during this stage to ensure robust seedling development.

Vegetative Growth

As the seedlings mature, they enter the vegetative growth phase, characterized by the rapid development of leaves and stems. During this stage, ample sunlight, proper nutrition, and well-draining soil are essential. Pruning may be employed to encourage bushier growth, promoting a fuller and more productive plant.

Flowering and Reproductive Stage

The transition to the flowering stage signals the onset of reproductive activities. At this point, the plant allocates energy towards flower and seed production. While flowering is a natural part of the life cycle, prolonged exposure to stressors such as extreme temperatures or nutrient deficiencies may impact overall yield. Adequate fertilization and environmental control are pivotal during this critical phase.

Harvesting

The final growth phase is harvesting, where the mature leaves are carefully picked to maximize quality and yield. Timing is crucial during this stage, as harvesting too early or too late can affect the taste, texture, and nutritional content of the leaves. Regular monitoring and observation are necessary to identify the optimal harvest window for leaf amaranth.

Monitoring Plant Health

To ensure a thriving leaf amaranth crop and maximize business potential, consistent monitoring of plant health is imperative. This involves a multi-faceted approach to detect and address issues promptly.

Visual Inspection

Regular visual inspections of the leaves, stems, and overall plant appearance are essential. Signs of discoloration, wilting, or abnormal growth patterns may indicate nutrient deficiencies, pests, or diseases. Early detection through visual inspection allows for timely intervention and mitigation.

Soil Analysis

Conducting periodic soil tests is a proactive measure to assess nutrient levels and soil composition. Adjusting fertilizer applications based on soil analysis results ensures that the plants receive the necessary nutrients for optimal growth. Maintaining the right pH level is equally important, as leaf amaranth thrives in slightly acidic to neutral soils.

Pest and Disease Management

Implementing integrated pest management strategies is crucial to ward off potential threats to plant health. Identify and address pest issues promptly, utilizing natural predators, organic pesticides, or cultural practices. Regularly monitor for common diseases such as powdery mildew or leaf spot and employ appropriate treatments if necessary.

Troubleshooting Common Issues

Even with meticulous care, leaf amaranth may encounter challenges that require swift resolution. Being equipped to troubleshoot common issues is integral to maintaining a robust and profitable cultivation operation.

Nutrient Deficiencies

Yellowing leaves or stunted growth may indicate nutrient deficiencies. Conduct a thorough nutrient analysis and adjust fertilizer applications accordingly. Common deficiencies include nitrogen, phosphorus, and iron, which can be rectified through targeted nutrient supplementation.

Pest Infestations

Common pests like aphids, mites, or caterpillars can quickly compromise plant health. Introduce natural predators, use organic pesticides, or implement physical barriers to deter pests. Regular monitoring is key to early detection and prevention of infestations.

Disease Management

Fungal or bacterial diseases can impact leaf amaranth, leading to reduced yields. Proper sanitation practices, well-ventilated growing conditions, and the application of fungicides or bactericides can help manage and prevent the spread of diseases.

Recording Growth Data

Maintaining detailed records of growth data is instrumental for informed decision-making and continuous improvement in leaf amaranth cultivation.

Growth Metrics

Record essential growth metrics such as plant height, leaf size, and stem diameter at regular intervals. These metrics provide insights into the overall development of the crop and help identify patterns or abnormalities.

Environmental Conditions

Document environmental factors like temperature, humidity, and light exposure. Understanding how these variables impact growth allows for optimization of cultivation practices and adaptation to seasonal changes.

Pest and Disease Incidences

Keep a log of pest and disease occurrences, noting the specific pests or diseases encountered, the severity of the infestation, and the measures taken for control. This information aids in refining pest management strategies and preventing recurrence.

Harvest Yields

Track harvest yields over time to assess the success of cultivation practices. This data is valuable for forecasting future yields, optimizing resource allocation, and making informed business decisions.

In conclusion, mastering the growth stages, monitoring plant health, troubleshooting common issues, and recording growth data are integral components of successfully growing leaf amaranth for business. By embracing a comprehensive and proactive approach, cultivators can ensure a consistent and high-quality supply of this valuable leafy green for the market.

CHAPTER EIGHT

HARVESTING TECHNIQUES

Determining Optimal Harvest Time

Leaf amaranth, a nutritious and fast-growing crop, demands meticulous attention to determine the optimal harvest time. Harvesting at the right moment ensures peak nutritional content and overall crop quality. The ideal time to harvest leaf amaranth is when the plants have reached full maturity, typically between 30 to 40 days after planting. Subtle cues such as leaf color, size, and texture can guide farmers in identifying the prime harvest window.

Leaf Color: One key indicator is the color of the leaves. Harvesting should be timed when the leaves exhibit a vibrant and deep green color. This indicates optimal chlorophyll content, a crucial component of the plant's nutritional value.

Leaf Size: Monitoring the size of the leaves is equally important. The leaves should be sufficiently large, ensuring a substantial yield. Smaller leaves may lack the desired nutrient concentration, affecting the overall quality of the harvest.

Texture and Consistency: Another factor to consider is the texture of the leaves. The foliage should feel crisp and tender to the touch, indicating peak freshness. The consistency in texture across the entire crop is vital for uniform quality.

Harvesting Methods

Various harvesting methods can be employed to efficiently gather leaf amaranth while preserving its nutritional integrity.

Selective Harvesting: This method involves carefully picking individual leaves from mature plants. It allows for a prolonged harvest period, ensuring a steady supply of fresh leaves. Selective harvesting is particularly beneficial for small-scale operations or when a continuous harvest is desired.

Cut-and-Come-Again Technique: The cut-and-come-again technique involves cutting the outer leaves of the plant, leaving the central leaves untouched. This encourages new growth, extending the overall harvest period. This method is sustainable and suitable for both small and large-scale cultivation.

Whole Plant Harvest: In some cases, especially for batch processing or when a quick harvest is needed, harvesting the entire plant at once is a viable option. However, this method requires careful planning to avoid disrupting the crop cycle and to ensure that subsequent planting can occur without delay.

Post-Harvest Handling

Post-harvest handling plays a crucial role in maintaining the freshness and quality of leaf amaranth. Proper handling practices are essential to prevent deterioration and ensure the marketability of the crop.

Immediate Cooling: After harvest, leaf amaranth should be promptly cooled to slow down the metabolic processes that lead to wilting and decay. Rapid cooling helps preserve the crispness and nutritional content of the leaves. Using cool storage facilities or refrigeration is recommended for larger harvests.

Cleaning and Sorting: Thorough cleaning and sorting are essential steps in post-harvest handling. Removing debris, damaged leaves, and any impurities enhances the visual appeal of the product and contributes to a longer shelf life.

Packaging for Freshness: Packaging plays a pivotal role in maintaining the freshness of leaf amaranth. Packaging materials should be breathable to prevent moisture build-up, which can lead to mold and decay. Vacuum-sealed bags or perforated containers are common choices.

Quality Control Measures

Quality control measures are indispensable in the leaf amaranth business to ensure that the final product meets market standards and consumer expectations.

58

Nutrient Analysis: Regular nutrient analysis of harvested samples helps in monitoring the nutritional content of the crop. This information is crucial for marketing and labeling the product accurately, providing consumers with valuable information about the nutritional benefits.

Pest and Disease Management: Implementing effective pest and disease management practices is vital for maintaining quality. Regular scouting for pests and diseases, coupled with timely interventions, ensures that the harvested leaves are free from contaminants and safe for consumption.

Hygicnc and Sanitation: Maintaining a high level of hygiene and sanitation throughout the harvesting and post-harvest processes is paramount. Clean and well-maintained equipment, storage facilities, and handling areas prevent contamination and contribute to a high-quality end product.

In conclusion, successful leaf amaranth cultivation for business involves a holistic approach from determining the optimal harvest time to implementing quality control measures. By adopting best practices in harvesting methods, post-harvest handling, and quality control, growers can ensure a consistent and high-quality supply of leaf amaranth for the market.

CHAPTER NINE

PROCESSING AND VALUE ADDITION

Processing Leaf Amaranth

Harvesting and Sorting

Processing leaf amaranth begins with careful harvesting to ensure the leaves are at their peak freshness and nutritional value. Harvesting should be done in the early morning or late afternoon when the plant is well-hydrated, and the temperature is cooler. It is crucial to select leaves that are young and tender, as they offer the best texture and flavor. After harvesting, the leaves should be sorted to remove any damaged or discolored ones.

Cleaning and Washing

To maintain the quality of leaf amaranth, thorough cleaning and washing are essential. Remove any dirt, debris, or insects from the leaves by gently washing them in cold water. A mild, natural disinfectant can be used to ensure the elimination of any potential contaminants. Proper hygiene during this stage is crucial to prevent the introduction of harmful microorganisms.

Drying Techniques

Drying leaf amaranth can be done through various methods, such as air drying or using dehydrators. The chosen method should retain the vibrant color and nutritional content of the leaves. Careful monitoring is necessary to prevent over-drying, which could lead to a loss of flavor and nutrients. Properly dried leaf amaranth can be stored for an extended period without compromising its quality.

Creating Value-Added Products

Amaranth Leaf Powder

One innovative way to add value to leaf amaranth is by creating a nutrient-packed powder. Dried leaves can be ground into a fine powder, preserving their nutritional content. This powder can be used as a versatile ingredient in smoothies, soups, or as a seasoning. Rich in vitamins and minerals, amaranth leaf powder is a convenient and nutritious addition to various dishes.

Amaranth-infused Oils

Extracting the essence of leaf amaranth through oils adds a gourmet touch to culinary products. Amaranth-infused oils not only capture the unique flavor of the leaves but also bring forth their health benefits. These oils can be used in salad dressings, sautés, or as a finishing touch to dishes, providing a distinctive and health-conscious option for consumers.

Amaranth Snack Bars

Capitalizing on the rising demand for healthy snacks, creating amaranth snack bars is a lucrative venture. By combining amaranth leaves with other wholesome ingredients, such as nuts, seeds, and dried fruits, producers can offer a convenient, on-the-go snack that is not only delicious but also packed with nutrients. This product appeals to health-conscious consumers looking for alternatives to traditional snacks.

Packaging and Branding
Sustainable Packaging

In the modern market, consumers are increasingly conscious of the environmental impact of packaging. Utilizing sustainable and eco-friendly packaging for amaranth products enhances the brand's image and attracts environmentally conscious consumers. Options like biodegradable bags or reusable containers not only reduce the ecological footprint but also contribute to the overall appeal of the product.

Brand Identity

Developing a strong brand identity is crucial for the success of Amaranth products. This includes a compelling logo, distinctive packaging design, and a memorable brand name. Communicating the health benefits, sustainability practices, and the uniqueness of the product through branding helps differentiate it in a competitive market. Consistent branding across all platforms fosters brand recognition and loyalty.

Marketing Strategies

Online Presence

Establishing a robust online presence is imperative for reaching a wider audience. Utilize social media platforms, e-commerce websites, and digital marketing to showcase the benefits of amaranth products. Engage with the audience through informative content, recipes, and customer testimonials to build trust and credibility.

Collaborations and Partnerships

Collaborating with chefs, nutritionists, and influencers can amplify marketing efforts. Partnerships can lead to creative recipes, cooking demonstrations, and endorsements, effectively promoting amaranth products to a broader audience.

Aligning with health and wellness brands or participating in relevant events can further enhance visibility and credibility.

Educational Campaigns

Educational campaigns highlighting the nutritional value and health benefits of amaranth products can create awareness and build consumer trust. Providing informative content through blogs, webinars, or workshops establishes the brand as an authority in the field. Emphasizing the versatility of amaranth in various cuisines encourages consumers to incorporate it into their diets.

In conclusion, processing and adding value to leaf amaranth for business involves meticulous steps in harvesting, cleaning, and drying. Creating innovative products, adopting sustainable packaging, building a strong brand identity, and implementing effective marketing strategies are key elements in establishing a successful venture in the growing leaf amaranth market.

CHAPTER TEN

FINANCIAL PLANNING AND BUDGETING
Cost Estimation for Growing Leaf Amaranth for Business

Growing leaf amaranth for business requires a comprehensive understanding of the associated costs. To establish a robust financial plan, one must consider various elements in the cost estimation process.

Land and Infrastructure:

The first significant expense in cultivating leaf amaranth is acquiring suitable land and setting up the necessary infrastructure. Costs may include land purchase or lease, irrigation systems, fencing, and other essential facilities. Additionally, investments in greenhouse technology for controlled cultivation can contribute to higher upfront costs but may yield long-term benefits.

Seeds and Planting Materials:

Choosing the right seeds and planting materials is crucial for a successful leaf amaranth cultivation venture. Costs related to acquiring high-quality seeds, seedlings, or saplings should be factored in. Moreover, investing in hybrid or genetically improved varieties may enhance yield potential, impacting overall profitability.

Labor and Management:

Skilled labor is vital for the cultivation of leaf amaranth. The costs associated with hiring skilled agricultural workers, farm managers, and other necessary personnel should be considered. Proper training programs and employee welfare measures may also contribute to the overall labor expenses.

Fertilizers and Crop Protection:

Ensuring the health and productivity of leaf amaranth plants requires a strategic approach to fertilization and pest control. Costs for acquiring fertilizers, pesticides, herbicides, and other crop protection measures should be estimated. Adopting sustainable and organic farming practices may influence these expenses.

Equipment and Machinery:

Investing in modern agricultural equipment and machinery can streamline farming operations and improve efficiency. Tractors, plows, harvesters, and other necessary tools contribute to upfront costs but may result in long-term savings through increased productivity.

Revenue Projections for Growing Leaf Amaranth for Business

Accurate revenue projections are essential for financial planning, providing insights into the potential profitability of growing leaf amaranth for business.

Market Analysis and Pricing Strategy:

Understanding the market demand for leaf amaranth and establishing a competitive pricing strategy is fundamental to revenue projections. Conducting market research to identify consumer preferences, trends, and potential competitors can inform pricing decisions and revenue forecasts.

Yield Estimates and Harvest Cycles:

Estimating the average yield per acre and considering the frequency of harvest cycles are critical components of revenue projections. Factors such as climate, soil quality, and cultivation practices can influence yield, and these should be carefully analyzed for accurate forecasting.

Sales Channels and Distribution:

Identifying the most effective sales channels and distribution methods is vital for revenue generation. Whether selling directly to consumers, through local markets or establishing partnerships with retailers, the chosen distribution strategy will impact overall revenue and market reach.

Value-Added Products:

Exploring opportunities for value addition can significantly enhance revenue streams. Processing leaf amaranth into value-added products such as salads, juices, or packaged goods can open new market avenues and increase profit margins.

Return on Investment Analysis for Growing Leaf Amaranth for Business

Analyzing the return on investment (ROI) is imperative to evaluate the financial viability of growing leaf amaranth for business.

Cost-Benefit Analysis:

Conducting a comprehensive cost-benefit analysis helps in understanding the relationship between the initial investment and the expected returns. This involves comparing the total costs incurred with the projected revenue to determine the profitability of the venture.

Break-Even Point Calculation:

Identifying the break-even point is crucial for understanding when the business will start generating profits. Calculating the point at which total revenue equals total costs provides insights into the timeline for achieving financial sustainability.

Risk Assessment:

Evaluating potential risks and uncertainties is integral to ROI analysis. Factors such as market fluctuations, weather conditions, and pest infestations can impact returns. Implementing risk mitigation strategies and considering different scenarios in the analysis enhances the overall reliability of ROI projections.

Funding Options for Growing Leaf Amaranth for Business

Securing appropriate funding is essential for successfully establishing and operating a leaf amaranth cultivation business.

Self-Financing:

Entrepreneurs may choose to fund the venture using personal savings or profits from other business endeavors. Self-financing provides independence and avoids the complexities associated with external funding sources.

Bank Loans and Financial Institutions:

Traditional bank loans and financial institutions offer a common avenue for obtaining funding. Entrepreneurs can present a detailed business plan, including financial projections, to secure loans for land acquisition, infrastructure development, and operational expenses.

Government Grants and Subsidies:

Many governments provide grants, subsidies, and financial incentives to promote agriculture and sustainable farming practices. Exploring these opportunities can significantly reduce the financial burden and enhance the feasibility of the leaf amaranth cultivation venture.

Investment from Private Investors:

Securing investments from private investors or venture capitalists is another option. Presenting a compelling business proposal and showcasing the potential for growth and profitability may attract external investors.

In conclusion, a comprehensive financial plan for growing leaf amaranth for business requires a meticulous analysis of costs, revenue projections, return on investment, and funding options. By considering these factors in detail, entrepreneurs can make informed decisions and set the stage for a successful and sustainable venture.

CHAPTER ELEVEN

CASE STUDIES

Successful Leaf Amaranth Businesses

Leaf amaranth, known for its nutritional value and versatility in culinary applications, has become a lucrative business venture for many entrepreneurs. This section explores the success stories of businesses that have flourished by growing leaf amaranth.

1. Organic Leaf Amaranth Farming

In the realm of successful leaf amaranth businesses, the emphasis on organic farming practices stands out. Farms that have embraced organic cultivation methods have not only met the increasing demand for pesticide-free produce but have also positioned themselves as premium suppliers in the market. By adopting sustainable and eco-friendly practices, these businesses have not only capitalized on the health-conscious consumer trend but have also contributed to environmental conservation.

2. Diversification of Products

Successful leaf amaranth businesses often go beyond traditional farming practices. They explore the potential of different products derived from leaf amaranth, such as packaged salads, smoothie blends, and even value-added products like amaranth-based snacks. Diversifying the product line not only taps into various consumer preferences but also ensures a year-round market presence, mitigating the impact of seasonal fluctuations.

3. Innovative Marketing Strategies

In the competitive world of agriculture and food products, successful leaf amaranth businesses leverage innovative marketing strategies to stand out. They engage in storytelling, emphasizing the journey from farm to table, highlighting the nutritional benefits of leaf amaranth, and even collaborating with chefs to create exclusive recipes. Social media platforms play a crucial role in creating brand awareness, with visually appealing content showcasing the vibrant green leaves and the farm-to-fork process.

Lessons Learned

The journey of growing leaf amaranth for business comes with its own set of challenges and lessons. Understanding these lessons is crucial for aspiring entrepreneurs to navigate the complexities of the industry.

1. Soil Quality Matters

One key lesson learned is the critical importance of soil quality. Leaf amaranth thrives in well-drained, fertile soil. Businesses that invest in soil testing and subsequent amendments tailored to the specific needs of leaf amaranth have reported higher yields and better-quality produce. Neglecting soil quality can lead to nutrient deficiencies, affecting both the growth and nutritional content of the crop.

2. Adaptability to Climate Conditions

Leaf amaranth is known for its adaptability to various climate conditions. However, businesses have learned the importance of closely monitoring weather patterns and adjusting cultivation practices accordingly. Whether dealing with excessive rainfall, drought, or temperature fluctuations, successful businesses employ strategies such as controlled environment agriculture or adjusting planting schedules to optimize yield despite changing climate conditions.

3. Supply Chain Management

Efficient supply chain management is a lesson that cannot be overstated. Businesses have learned that a well-organized supply chain, from harvest to distribution, is crucial for maintaining the freshness and quality of leaf amaranth. Implementing proper harvesting techniques, post-harvest handling, and streamlined transportation methods ensures that the produce reaches consumers in optimal condition, leading to customer satisfaction and repeat business.

Best Practices

Achieving success in growing leaf amaranth for business involves implementing best practices that cover every aspect of cultivation and marketing.

1. Integrated Pest Management (IPM)

Adopting an Integrated Pest Management (IPM) approach is a best practice that minimizes the use of chemical pesticides. Businesses that successfully incorporate biological controls, crop rotation, and companion planting reduce the impact of pests while maintaining the integrity of the crop. This not only aligns with consumer preferences for pesticide-free produce but also contributes to sustainable farming practices.

2. Technology Integration for Precision Agriculture

Embracing technology for precision agriculture is a best practice that enhances efficiency and productivity. Businesses invest in tools like soil sensors, drones, and data analytics to monitor crop health, optimize irrigation, and predict potential issues. This proactive approach not only ensures a higher yield but also minimizes resource usage, aligning with sustainable and cost-effective farming practices.

3. Building Partnerships in the Culinary Industry

Forming partnerships with chefs, restaurants, and food influencers is a best practice that extends the reach of leaf amaranth businesses. By collaborating on exclusive recipes, participating in food events, and featuring in culinary shows, businesses can create a buzz around their products and tap into new markets. Building these relationships not only boosts sales but also elevates the brand image of leaf amaranth as a versatile and desirable ingredient.

In conclusion, the success of growing leaf amaranth for business lies in a combination of organic farming, diversification, innovative marketing, and the implementation of lessons learned and best practices. Entrepreneurs in this field must stay adaptive, continuously learn from their experiences, and embrace the best practices that pave the way for a sustainable and profitable venture.

CHAPTER TWELVE

FUTURE TRENDS AND INNOVATIONS
Emerging Technologies in Growing Leaf Amaranth for Business

Precision Agriculture and IoT Integration: Precision agriculture is playing a pivotal role in modernizing leaf amaranth cultivation. Integration with the Internet of Things (IoT) allows farmers to monitor and control various parameters such as soil moisture, temperature, and nutrient levels in real time. Sensor-based technologies and automated systems enhance efficiency, reduce resource wastage, and contribute to higher yields. As these technologies continue to advance, businesses involved in growing leaf amaranth can leverage them to optimize production processes and enhance overall crop quality.

Biotechnology for Crop Improvement: Biotechnological advancements are revolutionizing the leaf amaranth industry. Genetic engineering techniques allow for the development of amaranth varieties with improved resistance to diseases, pests, and environmental stressors. Moreover, the modification of nutritional content can result in varieties that are not only more robust but also nutritionally enhanced. Businesses can explore partnerships with biotechnology firms to access cutting-edge genetic solutions, leading to the production of superior leaf amaranth varieties that meet evolving market demands.

Sustainable Practices in Growing Leaf Amaranth for Business

Organic Farming and Certification: As consumer preferences shift towards healthier and sustainably produced food, the demand for organically grown leaf amaranth is on the rise. Adopting organic farming practices not only aligns with environmental sustainability but also opens up premium markets for businesses. Obtaining organic certifications assures consumers of the product's purity and adherence to ethical farming practices. Implementing organic farming methods, such as composting, crop rotation, and natural pest control, can enhance the marketability of leaf amaranth products.

Water-Efficient Cultivation Methods: Water scarcity is a growing concern globally, and agriculture is under pressure to adopt water-efficient practices. In the context of leaf amaranth cultivation, businesses can explore hydroponics and aeroponics – soilless cultivation methods that use significantly less water compared to traditional farming. These techniques not only conserve water but also provide controlled environments for optimal growth, resulting in higher yields. Implementing water-efficient technologies aligns with sustainable agriculture practices and positions businesses as environmentally responsible players in the market.

Market Trends in Growing Leaf Amaranth for Business

Rising Consumer Awareness and Demand: Consumers are becoming increasingly health-conscious, driving the demand for nutrient-dense and plant-based foods. Leaf amaranth, being a rich source of vitamins, minerals, and antioxidants, is gaining popularity among health-conscious consumers. Businesses can capitalize on this trend by promoting the nutritional benefits of leaf amaranth and incorporating it into various food products. The rising demand for organic and locally sourced produce further creates opportunities for businesses to establish a strong market presence.

Innovation in Product Development: The market for leaf amaranth is evolving beyond traditional fresh produce. There is a growing trend in developing innovative products such as amaranth-based snacks, beverages, and value-added food items. Businesses can diversify their product offerings to cater to changing consumer preferences and tap into the expanding market for healthy and convenient food options. Collaborating with food technologists and chefs can lead to the creation of unique and appealing leaf amaranth products that stand out in a competitive market.

Opportunities for Growth in Growing Leaf Amaranth for Business

Export Market Expansion: With the increasing global awareness of the nutritional benefits of leaf amaranth, there is a significant opportunity for businesses to tap into international markets. Establishing strategic partnerships with distributors and retailers in other regions can open up avenues for export. Adhering to international quality standards and certifications is crucial to gain trust in foreign markets and ensure the success of export initiatives.

Value Chain Integration: Businesses can explore opportunities for vertical integration within the leaf amaranth value chain. This involves participating in multiple stages of production, processing, and distribution. By integrating backward into seed production or forward into product development and marketing, companies can gain better control over quality, reduce dependency on external suppliers, and create new revenue streams. Value chain integration also allows businesses to differentiate themselves by offering comprehensive solutions, from farm to consumer.

In conclusion, the future of growing leaf amaranth for business is shaped by a combination of emerging technologies, sustainable practices, market trends, and strategic growth opportunities. Embracing innovation, staying attuned to consumer preferences, and adopting environmentally friendly practices will position businesses for success in this dynamic and evolving industry.

www.ingramcontent.com/pod-product-compliance
Lightning Source LLC
Chambersburg PA
CBHW070757290526
45795CB00002B/585